# 2nd International Conference on Plant Synthetic Biology, Bioengineering and Biotechnology 2018



Clearwater, Florida, USA
29 November – 1 December 2018

ISBN: 978-1-5108-9679-6

**Printed from e-media with permission by:**

Curran Associates, Inc.
57 Morehouse Lane
Red Hook, NY 12571

**Some format issues inherent in the e-media version may also appear in this print version.**



Printed with permission by Curran Associates, Inc. (2019)

For permission requests, please contact AIChE
at the address below.

AIChE
120 Wall Street, FL 23
New York, NY 10005-4020

Phone: (800) 242-4363
Fax: (203) 775-5177

www.aiche.org

**Additional copies of this publication are available from:**

Curran Associates, Inc.
57 Morehouse Lane
Red Hook, NY 12571 USA
Phone: 845-758-0400
Fax: 845-758-2633
Email: curran@proceedings.com
Web: www.proceedings.com

# 2nd International Conference on Plant Synthetic Biology, Bioengineering, and Biotechnology proceedings

## General Submissions

# New Tools for an Old Problem

**Conference:** International Conference on Plant Synthetic Biology and Bioengineering

**Year:** 2018

**Proceeding:** 2nd International Conference on Plant Synthetic Biology, Bioengineering, and Biotechnology

**Group:** General Submissions

**Session:** Tissue Culture and Transformation of Recalcitrant Species

**Time:** Thursday, November 29, 2018 - 12:25pm-12:50pm

## Authors:

Boutilier, K., Wageningen University and Research

Somatic and gametophytic embryogenesis are two examples of induced plant cell totipotency, where embryos develop from vegetative or gametophytic tissues, rather than from the zygote after fertilisation. These two types of *in vitro* embryogenesis form the basis for a number of plant breeding and biotechnology applications, including clonal propagation, doubled-haploid production and regeneration after transformation, but also offer good model systems for understanding how plant cells are reprogrammed to follow a new developmental pathway. The classical tissue culture approach used to overcome recalcitrance for*in vitro* regeneration in crops is to identify empirically the explant and culture parameters that contribute to efficient production systems. Almost all of the major breakthroughs in plant tissue culture have been achieved in this way, and although successful, this approach is often time consuming and inefficient, as only a few parameters can be tested at one time. A molecular-genetic approach is often used to identify the molecular pathways driving plant cell totipotency in model plant systems, but can be confounded by genetic redundancy, lethality, or pleiotropy.

Our lab is using a third approach, chemical screening, to overcome recalcitrance for *in vitro* embryogenesis in crops, and to better understand the mechanistic basis for induced totipotency. In this approach, individual compounds from commercial targeted and non-targeted compound libraries are screened for their ability to enhance embryogenesis, while keeping the tissue culture conditions constant. I will highlight the general concepts behind chemical screening and demonstate how we have used this approach to identify competence factors for gametophytic and somatic embryogenesis.

# Maize Transformation at Pioneer - Meeting Future Demands for Genome Modification

**Conference:** International Conference on Plant Synthetic Biology and Bioengineering

**Year:** 2018

**Proceeding:** 2nd International Conference on Plant Synthetic Biology, Bioengineering, and Biotechnology

**Group:** General Submissions

**Session:** Tissue Culture and Transformation of Recalcitrant Species

**Time:** Thursday, November 29, 2018 - 12:50pm-1:15pm

## Authors:


Gordon-Kamm, B., Corteva Agriscience Agricultural Division of DowDuPont
Lowe, K., Corteva Agriscience Agricultural Division of DowDuPont
George, H., Corteva Agriscience Agricultural Division of DowDuPont
Wang, N., Corteva Agriscience Agricultural Division of DowDuPont
Anand, A., Corteva Agriscience Agricultural Division of DowDuPont
Wu, E., Corteva Agriscience Agricultural Division of DowDuPont
Arling, M., Corteva Agriscience Agricultural Division of DowDuPont
Jones, T., Corteva Agriscience Agricultural Division of DowDuPont



Although cereal transformation has continued to improve over the years, the process has generally remained constrained to a few genotypes per crop, and methods are still slow and labor intensive. The overall impact has been that cereal transformation, genome modifications and high-throughput transformation remain beyond the reach of almost all academic labs.

Recent progress in our labs is rapidly changing this situation for cereals. By focusing on the overexpression of the maize Babyboom (BBM) and Wuschel2 (WUS2) genes, we can routinely produce high transformation frequencies in numerous previously non-transformable maize inbreds. Recent optimization of the promoters driving BBM and WUS2 expression has resulted in direct germination of somatic embryos to recover transgenic plants (eliminating all callus steps), making maize inbreds such as B73 and Mo17 easily transformable using Agrobacterium strain LBA4404. Of even greater import to genome modification methods, this process is largely genotype independent and regenerated plants can be sent to the greenhouse in less than half the time of conventional methods. Another limitation for many monocots is the intensive labor and greenhouse space required to supply immature embryos for transformation. As a new alternative to immature embryos, we use BBM and WUS2 to recover transgenic events directly from either embryo slices from mature seed or leaf segments from seedlings in a variety of Pioneer inbreds, routinely recovering healthy, fertile T0 plants. We will also describe a new system for Agrobacterium-mediated co-transformation that greatly simplifies the process from vector construction through analysis.

# Improved Plant Synthetic Biology Using a Versatile and Robust Agrobacterium-Based Gene Stacking System

**Conference:** International Conference on Plant Synthetic Biology and Bioengineering

**Year:** 2018

**Proceeding:** 2nd International Conference on Plant Synthetic Biology, Bioengineering, and Biotechnology

**Group:** General Submissions

**Session:** Synthetic Biology Tools and Quantitative Aspects of Gene Circuit Design

**Time:** Thursday, November 29, 2018 - 3:20am-3:45am

## Authors:

Thilmony, R., USDA-ARS

Plant synthetic biology provides a means for the rapid genetic improvement of crops and will enable future improvements of complex traits like yield and nutritional quality through the introduction and coordinated expression of multiple genes. GAANTRY (Gene Assembly in*Agrobacterium* by Nucleic acid Transfer using Recombinase technologY) is a flexible and effective system for stably stacking multiple genes within an *Agrobacterium* virulence plasmid Transfer-DNA (T-DNA) (Collier et al. The Plant Journal 2018). The system utilizes unidirectional site-specific recombinases *in vivo* and an alternating selection scheme to sequentially assemble multiple genes into a single transformation construct. To demonstrate GAANTRYâ€™s capabilities, 10 cargo sequences were sequentially stacked together to produce a 28.5 kilobase pair T-DNA, which was used to generate hundreds of transgenic*Arabidopsis* events. Approximately 90% of the events identified using the dual antibiotic selection screen exhibited all of the introduced traits. A total of 68% of the tested lines carried a single copy of the selection marker transgene located near the T-DNA left border and only 8% contained sequence from outside the T-DNA. Research is also underway to demonstrate that the GAANTRY system can successfully genetically engineer rice, as well as other important crop species. Thus, GAANTRY is a powerful, yet simple to use, new tool for transgene stacking and plant synthetic biology.

# Parts-Prospecting for the Methanol Economy: Yhbo and Formaldehyde Resilience

**Conference:** International Conference on Plant Synthetic Biology and Bioengineering

**Year:** 2018

**Proceeding:** 2nd International Conference on Plant Synthetic Biology, Bioengineering, and Biotechnology

**Group:** General Submissions

**Session:** Synthetic Biology Tools and Quantitative Aspects of Gene Circuit Design

**Time:** Thursday, November 29, 2018 - 12:00pm-12:25pm

## Authors:

Patterson, J. A., University of Florida
He, H., Max Planck Institute of Molecular Plant Physiology
Bar-Even, A., Max Planck Institute of Molecular Plant Physiology
Wilson, M. A., University of Nebraska
Hanson, A. D., University of Florida

Formaldehyde (HCHO) is a highly reactive carbonyl compound that cross-links and formylates proteins, DNA, and small molecules, making it both cytotoxic and genotoxic. The mechanism of HCHO damage, and the enzyme systems that repair or prevent it, are critical to natural and synthetic methylotrophy â€“ and hence to the proposed methanol economy. Methanol produced in various ways is a potential cheap and sustainable carbon source for biofuel production and other SynBio applications. As the oxidation of methanol to HCHO is the first step in methanol metabolism, HCHO production is an unavoidable facet of methylotrophy. The accumulation of

HCHO can damage the metabolic machinery in platform cells and increase their mutation rate, leading to rapid loss of engineered production traits. Little is known about whether or how cells repair HCHO-damage to proteins, DNA, and small molecules. Comparative genomic analysis indicates that the enzyme YhbO is associated with HCHO-related proteins. YhbO belongs to the DJ-1 protein family, which is implicated in carbonyl damage repair. Deletion ofyhbO in Escherichia coli strain BW25113 increases sensitivity to exogenous HCHO; this phenotype is specific to HCHO and is independent of carbon source. Interestingly, deleting yhbO in other E. coli K12 backgrounds (e.g., MG1655) does not lead to HCHO-sensitivity. YhbO thus has potential as a damage-repair part in engineering synthetic methylotrophy. We are now testing possible mechanisms by which YhbO confers HCHO resilience, (e.g., by deformylating proteins) using genetic and biochemical approaches and investigating the gene(s) responsible for the difference between the BW25113 and MG1655 strains.

# A Mathematical Model in Absolute Units for the Arabidopsis Circadian Oscillator

**Conference:** International Conference on Plant Synthetic Biology and Bioengineering

**Year:** 2018

**Proceeding:** 2nd International Conference on Plant Synthetic Biology, Bioengineering, and Biotechnology

**Group:** General Submissions

**Session:** Synthetic Biology Tools and Quantitative Aspects of Gene Circuit Design

**Time:** Thursday, November 29, 2018 - 12:25pm-12:50pm

## Authors:

Urquiza-García, U., The University of Edinburgh
Millar, A. J., The University of Edinburgh

*Arabidopsis thaliana* presents circadian oscillations in more than 30% of its transcriptome. These dynamics are generated by a biochemical oscillator made of a collection of interlocked feedback loops. Mathematical modeling has been fundamental for understanding how circadian rhythms emerge form molecular interactions. However, up to now, models have lacked absolute units on the mass scale. The advent of genome editing tools requires a new type of modeling were DNA sequences inform the parameter values in models and how changes in sequence impact the dynamics of genetic circuits. Here we present how absolute quantifications of clock transcription factors using a state of the art bioluminescent reporter combined with biophysical models of transcription factor binding is a first step for incorporating sequence information into dynamical models. We provide a framework for exploring the possible implications of editing ciselements on the dynamics of the *Arabidopsis* circadian clock.

# Engineering Root Architecture with New Tools for Controlling Gene Expression in Plants

**Conference:** International Conference on Plant Synthetic Biology and Bioengineering

**Year:** 2018

**Proceeding:** 2nd International Conference on Plant Synthetic Biology, Bioengineering, and Biotechnology

**Group:** General Submissions

**Session:** Synthetic Biology Tools and Quantitative Aspects of Gene Circuit Design

**Time:** Thursday, November 29, 2018 - 1:15pm-1:40pm

## Authors:

Brophy, J. A. N., Carnegie Institution of Washington
Dinneny, J. R., Carnegie Institution of Washington

The shape of a plantâ€™s root system influences its ability to reach essential nutrients in the soil or to acquire water during drought. Progress in engineering plant roots to optimize water and nutrient acquisition has been limited by our capacity to design and build genetic programs that alter root growth in a predictable manner. We are building a library of synthetic transcription factors and using them to construct gene circuits that reprogram root development. The genetic circuits will be used to express mutant auxin-response transcription factors (AUX/IAAs) in specific root cell-types to alter root structure. In addition to synthetic gene circuits, we are constructing a platform for site specific integration of transgenes in Arabidopsis. This site specific integration platform should accelerate the development of genetic circuits in plants by reducing the variability in gene expression produced by random transgene integration and shortening the design-build-test cycle. We are currently testing methods of preventing random T-DNA integration after Agrobacterium-mediated DNA delivery. If successful and efficient, our platform will direct T-DNAs to integrate a specific location within the plant genome. This will accelerate the pace of plant research by generating clonal plant lines that can be phenotyped in the T1 generation.

# In silico Model for Mining the Cis-Regulatory Determinants of Tissue-Specific Gene Expression

**Conference:** International Conference on Plant Synthetic Biology and Bioengineering

**Year:** 2018

**Proceeding:** 2nd International Conference on Plant Synthetic Biology, Bioengineering, and Biotechnology

**Group:** General Submissions

**Session:** Synthetic Biology Tools and Quantitative Aspects of Gene Circuit Design

**Time:** Thursday, November 29, 2018 - 2:05pm-2:30pm

## Authors:

Megraw, M., Oregon State University
Fraser, V. N., Oregon State University
Ansariola, M., Oregon State University
O'Neil, S., Oregon State University
Filichkin, S., Oregon State University

Gene expression across tissues is regulated by an unknown number of determinants, including prevalence of transcription factors (TFs) and their binding sites along with other aspects of cellular state. Recent studies have suggested the importance of both genetic and epigenetic aspects, at least two of which have substantial literature support as causal determinants of tissue specificity: TF binding sites, and chromatin accessibility at those sites. In order to investigate the extent and relative contributions of these potential determinants, we produced three datasets for both leaf and root tissues of Arabidopsis thaliana plants: TSS-seq data to identify Transcription Start Sites, OC-seq data to identify regions of Open Chromatin, and RNA-seq data to assess gene expression levels. For those genes that are differentially expressed between root and leaf, we constructed a model incorporating chromatin accessibility with TF binding information upstream of TSS locations, with the goal of predicting the tissue in which each of these genes would be upregulated. The resulting model was highly accurate when both chromatin structure and sequence were considered (over 90% auROC and auPRC), allowing one to predict the tissue in which a given gene will express. Specifically, one can use the model to (1) create â€œin silico knockoutsâ€ of TFs that strongly influence the predicted tissue of expression, and (2) identify collections of TFs whose absence moves a specific native gene promoter across the decision boundary to predict expression in a different tissue. The ultimate goal of this work is to develop a rational design process for the tissue-specific targeting of plant gene promoters.

# Development of Novel Optogenetic Tools By Continuous Directed Evolution of Plant Blue-Light-Dependent Interactions

**Conference:** International Conference on Plant Synthetic Biology and Bioengineering

**Year:** 2018

**Proceeding:** 2nd International Conference on Plant Synthetic Biology, Bioengineering, and Biotechnology

**Group:** General Submissions

**Session:** Synthetic Biology Tools and Quantitative Aspects of Gene Circuit Design

**Time:** Thursday, November 29, 2018 - 2:55pm-3:20am

## Authors:

Liu, H., UCLA

Arabidopsis thaliana Cryptochrome2 (AtCRY2) is a blue-light receptor that mainly regulates plant photomorphogenesis through blue-light-specific interactions with numerous protein partners. Such blue-light-specific interactions have been exploited in optogenetics to manipulated biological events in a timely and precisely manner. AtCRY2-AtCIB1 photodimerizer is one of the best characterized protein pairs for optical control of protein interactions. However, the application of CRY2-CIB1 was impeded by concomitant homodimerization of AtCRY2, and limited dynamic range. To overcome the limitations, this study focused on the development of a novel pair of blue-light-dependent interacting proteins: AtCRY2-AtBIC1. AtBIC1 (Blue-light Inhibitor of Cryptochromes 1) was recently identified as an endogenous inhibitor of AtCRY2. BIC1 interacts with CRY2 and inhibits homodimerization of CRY2 in the blue light. Additionally, the size of BIC1 is very small (only 15 kDa), which could allow gene fusion with minimal impact. We used a novel continuous directed evolution method developed by David Liuâ€™s lab, PACE (Phage Assisted Continuous Evolution), to evolve AtCRY2 to increase its interaction affinity with BIC1 in the blue light. The PACE method coupled random mutagenesis with selection, enabling dozens of rounds of automatic evolution in a single day with minimal human efforts. Using the method, we successfully isolated mutants of CRY2 with stronger interactions with BIC1. To increase the dynamic range and signal-to-background ratio of CRY2-BIC1 dimerizer, a counter evolution method was proposed to reduce background in darkness. Our study could provide improved tools for optical control of protein interactions.

# Opportunities for Plant Biotechnology to Advance Plant-Based Meat and Clean Meat

**Conference:** International Conference on Plant Synthetic Biology and Bioengineering

**Year:** 2018

**Proceeding:** 2nd International Conference on Plant Synthetic Biology, Bioengineering, and Biotechnology

**Group:** General Submissions

**Session:** Plant Foundry: Biopolymers, Biopharmaceuticals and Medically Effective Secondary Metabolites (Vaccines, Antibodies, and Metabolites)

**Time:** Friday, November 30, 2018 - 1:15pm-1:40pm

## Authors:

Specht, L., The Good Food Institute

Plant engineering and biotechnology present tremendous opportunities to improve the functionality of diverse plant-based ingredients for incorporation into alternatives to animal products. It is well known that animal agriculture poses massive problems for public health, the environment, and animal welfare, yet plant-based meat currently comprises only 0.3% of the U.S. meat market. Greater innovation in plant-based food technology is needed to develop alternatives that compete with animal products on taste, price, convenience, and functionality. Most plant-based products currently on the market are made primarily of soy protein or wheat protein, while hundreds of other plant protein sources are virtually unexplored for their functionality within plant-based meat formulations. This is due in part to a lack of targeted breeding or engineering efforts for these crops, which reduces their attractiveness to food scientists for exploration as novel protein ingredients in plant-based meat, egg, or dairy formulations. This talk will discuss how plant engineering approaches can improve upon these diverse plant protein cropsâ€™ functional traits such as palatability, protein quality, and structuring capability to enhance their performance in plant-based meat, egg, or dairy alternatives. Furthermore, the emerging field of clean meat â€“ producing genuine animal meat by directly cultivating animal cells in culture â€“ is exploring plant-based ingredients to fabricate scaffolding material for complex 3D tissues as well as to provide low-cost nutrients to support cell growth. The clean meat field can benefit from plant biotechnology approaches to produce these materials at large scale, low cost, and with tailored functional properties for applications in clean meat.

# Enhancing Photosynthesis, Vitamin E Production and Tolerance to Herbicides By Targeting Homogentisate Catabolism Using CRISPR/Cas9 Genome Editing System in Soybean

**Conference:** International Conference on Plant Synthetic Biology and Bioengineering

**Year:** 2018

**Proceeding:** 2nd International Conference on Plant Synthetic Biology, Bioengineering, and Biotechnology

**Group:** General Submissions

**Session:** Genome Editing for Crop Improvement

**Time:** Friday, November 30, 2018 - 12:00pm-12:25pm

## Authors:

Nguyen, C., University of Missouri
Cahoon, R., University of Nebraska
Nguyen, H., University of Nebraska
Do, P., University of Missouri
Saito, S., University of Nebraska
Stacey, G., University of Missouri
Clemente, T., University of Nebraska
Cahoon, E., University of Nebraska
Stacey, M., University of Missouri

Soybean (*Glycine max*) is a major plant source of protein and oil and produces important secondary metabolites beneficial for human health. In a previous study, we identified and characterized the function of homogentisate dioxygenase (HGO), which catalyzes the committed enzymatic step in homogentisate catabolism, through a forward genetic screen screen of a soybean fast neutron (FN) mutant population. The soybean fast neutron mutant line, MO12, produced brown seeds with higher concentrations of homogentisate and Vitamin E in seeds. In this study, we used CRISPR/Cas9 genome editing system to further study the effect of *HGO* genes on the homogentisate metabolism pathway in soybean. The dual gRNA CRISPR/Cas9 system was utilized to generate single and double *hgo1 hgo2* mutants of soybean. The *homozygous hgo1*mutant plants are phenocopies of the MO12 mutant line, whereas the*hgo2* mutant plants showed a similar phenotype to the wildtype, *Cv*. Throne. The homogentisate content in the seed of the double *hgo1 hgo2*mutant line was comparable to that of the *hgo1* mutant, indicating that GmHGO2 has no significant contribution to homogentisate catabolism in soybean seeds. However, we observed improved growth performance of the double mutant in the field. The double mutant line showed increased tolerance to HPPD-inhibiting herbicide, which interferes with homogentisate biosynthesis, and an increase in photosynthetic rate compared to wild type and single mutants. Overall, our results showed that GmHGO1 is the major isoform responsible for homogentisate catabolism in seeds, but both GmHGO1 and GmHGO2 contribute to homogentisate turn-over in vegetative tissues. Moreover, our results show that targeting homogentisate catabolism is an effective approach to improving soybean agronomic traits and that CRISPR/Cas9 system is a powerful tool for plant metabolic engineering in soybean.

# Engineering the Chloroplast Genome with Synthetic Biology Tools for Microcompartment Construction

**Conference:** International Conference on Plant Synthetic Biology and Bioengineering

**Year:** 2018

**Proceeding:** 2nd International Conference on Plant Synthetic Biology, Bioengineering, and Biotechnology

**Group:** General Submissions

**Session:** Engineering of the Chloroplast Genome

**Time:** Friday, November 30, 2018 - 12:00pm-12:25pm

## Authors:

Chaudhari, V., Cornell University
Hanson, M., Cornell University

Photosynthetic efficiency of C3 plants suffers from the slow catalytic rate and the reaction of ribulose 1,5-bisphosphate carboxylase/oxygenase (Rubisco) with $O_2$ instead of $CO_2$, leading to the costly process of photorespiration. Cyanobacteria are able to utilize a form of Rubisco exhibiting a higher catalytic rate by encapsulating the enzyme within a microcompartment known as a carboxysome. Engineering the cyanobacterial $CO_2$ concentrating mechanism (CCM) into chloroplasts is an approach to enhance photosynthesis in important C3 plants such as soybean, poplar, and Brassica species. Microcompartments may also be used to compartmentalize other biochemical reactions to confer new capabilities on transgenic plants.

Achieving such a goal requires improving the technology for expressing multiple proteins from chloroplast operons. We have been developing strategies and vectors for expressing as many as 12 cyanobacterial genes from the chloroplast genome. Our strategy uses parts flanked by two restriction sites on each side where the distal site is a Type-II rare cutter (8bp recognition) and proximal site is TypeIIS (S for shifted cleavage). Usage of Type-IIS restriction enzyme, which cuts outside the recognition site and produces non-palindromic overhang, brings several unique advantages to traditional BioBrick strategy while maintaining its features. First, more than two parts can be inserted in a vector backbone in a single round of cloning. Second, usage of *LguI* allows to make a â€˜scar-lessâ€™ 5â€™UTR-CDS junction or an in-frame joining of protein domains with 1-substitution (as opposed to two in the latest Biobrick standards). Third, the flexibility in the overhang sequence creates advantages similar to Golden Gate cloning. The strategy allows designing of case-specific approaches which can assemble even more number of parts simultaneously. And lastly, it reduces the domestication requirement for parts as it utilizes longer recognition sites for 3 out of 4 of the enzymes.

# Engineered RNA-Binding Protein for Gene Regulation in Non-Green Plastids

**Conference:** International Conference on Plant Synthetic Biology and Bioengineering

**Year:** 2018

**Proceeding:** 2nd International Conference on Plant Synthetic Biology, Bioengineering, and Biotechnology

**Group:** General Submissions

**Session:** Engineering of the Chloroplast Genome

**Time:** Friday, November 30, 2018 - 12:25pm-12:50pm

## Authors:

Maliga, P., Rutgers University
Yu, Q., Rutgers University
Barkan, A., University of Oregon

Non-green plastids are desirable for the expression of recombinant proteins in edible plant parts to enhance the nutritional value of tubers or fruits. However, plastid transgenes are expressed at extremely low levels in the amyloplasts of storage organs such tubers. Here we report a two-component regulatory system consisting of an engineered PPR10$^{GG}$ RNA binding protein and a cognate BS$^{GG}$ binding site upstream of a GFP reporter gene in the plastid genome. The BS$^{GG}$ binding site is not recognized by the resident potato PPR10 protein, restricting GFP protein accumulation to low levels in leaves. When the engineered PPR10$^{GG}$protein is expressed from the tuber-specific patatin promoter, GFP accumulation is enhanced 20-fold, from 0.06% to about 1.3% of total soluble protein in tubers while having little impact on GFP accumulation in leaves. The two-component regulatory system enables high-level transgene expression in non-photosynthetic plastids without interfering with chloroplast gene expression in leaves.

# Advances in Synthetic Biology Approaches for Production of High Value Products in Chloroplasts

**Conference:** International Conference on Plant Synthetic Biology and Bioengineering

**Year:** 2018

**Proceeding:** 2nd International Conference on Plant Synthetic Biology, Bioengineering, and Biotechnology

**Group:** General Submissions

**Session:** Engineering of the Chloroplast Genome

**Time:** Friday, November 30, 2018 - 12:50pm-1:15pm

## Authors:

Daniell, H., University of Pennsylvania

Chloroplasts are ideal bioreactors for production of proteins but require synthetic biology approaches to express eukaryotic genes utilizing prokaryotic protein synthetic machinery. A codon optimization algorithm has been developed utilizing highly expressed chloroplast genes from 133 sequenced chloroplast genomes using Java (Plant Physiology 172:62). High level expression of the largest human protein (891 kDa) â€" pentameric form of human Blood Clotting Factor VIII was demonstrated using this synthetic approach (Plant Biotechnology Journal 16: 1148). In addition, selectable marker genes have been removed to facilitate regulatory approval after expression of therapeutic, food or feed enzymes or proteins in lettuce chloroplasts. Recent advances in this field including commercial scale production of human therapeutic proteins in

FDA-approved cGMP facilities (Biomaterials 70: 84-93), development of tags to deliver protein drugs to targeted human cells or tissues (Biomaterials 80: 68-79), methods to quantify in planta drug dose using proteomic quantitation by parallel reaction monitoring analysis (Plant Physiology 172:62-77), long-term stability of proteins/enzymes at ambient temperature (Molecular Therapy, 24: 1342-1350), testing human drug doses in large animals (Molecular Therapy 25: 512-522), toxicology, pharmacokinetic and pharmacodynamics studies to obtain regulatory approval will be presented (Annual Review of Genetics, 50: 595-618; Genome Biology 17:134).

Five newly launched leaf-products will be compared here with 23 commercial microbial-enzyme products for textile, detergent or juice industries. Crude leaf-extract enzymes are functional at low concentrations without protease inhibitors. Contact-angle water droplet absorption by the FAMAS bioscouring videos exceeds 3-second industry requirements. Leaf-lipase/mannanase crude-extracts remove chocolate/mustard oil stains more efficiently at 70°C than commercial enzymes (<10% activity). Endo/exoglucanase crude leaf-extracts remove dye efficiently from denim surface and de-pilled knitted fabric. Leaf-pectinase powder efficiently clarified orange juice pulp. Thus, leaf-production platform offers a novel low-cost approach by elimination of fermentation, purification, concentration, formulation and cold-chain to revolutionize healthcare and enzyme industries.

# Combining Parts: Coupling Photosynthetic Electron Transport to Metabolic Engineering

**Conference:** International Conference on Plant Synthetic Biology and Bioengineering

**Year:** 2018

**Proceeding:** 2nd International Conference on Plant Synthetic Biology, Bioengineering, and Biotechnology

**Group:** General Submissions

**Session:** Engineering of the Chloroplast Genome

**Time:** Friday, November 30, 2018 - 1:15pm-1:40pm

## Authors:

Jensen, P. E., University of Copenhagen

Photosynthesis drives the production of ATP and NADPH mainly used to fix $CO_2$. Excess of redox power can be exploited for production of high-value compounds. Important natural products are often synthesized in low quantities by their host organism and can be difficult to produce by chemical synthesis because of their complex chemical structures. The cytochromes P450 (P450s) situated in the endoplasmic reticulum play key roles in natural product biosynthesis, and are powered by electron transfers from NADPH. We have shown that plant P450s expressed in chloroplasts and cyanobacteria are directed to the thylakoid membrane[1,2]. Here, the photosynthetic electron transport will support P450 catalytic activity independent of NADPH and dedicated reductases. In order to route reducing power more efficiently to P450s, we have fused them with ferredoxin (Fd) or flavodoxin-like FMN domains[3]. These fusions allow the P450s to obtain electrons for catalysis directly from the photosynthetic electron transport chain by interacting with photosystem I and make them competitive with the natural occurring ferredoxin requiring enzymes. Further dedicated redirection of reducing power can be obtained by scaffolding all the enzymes of a pathway on the thylakoid membrane. In a novel strategy, we have fused enzymes with transmembrane domains of TatB and TatC from the chloroplast twin arginine translocation system[4]. This reduced the accumulation of unwanted intermediates, side products and increased the accumulation of the end product fivefold. This work shows that chloroplasts and cyanobacteria are attractive for metabolic engineering, and suggests unexplored potential for engineering of photosynthetic electron transfer chains to accommodate heterologous enzymes.

[1] Gnanasekaran, T *et al.* (2016) J. Exp. Bot. 67(8):2495-506.

[2] Wlodarczyka, A *et al.* (2016) Met. Engineering 33: 1-11.

[3] Mellor, S *et al.* (2016) ACS Chem. Biol. 11(7):1862-9.

[4] Henriques de Jesus et al. (2017). Met. Engineering 44: 108-116.

# Directed Evolution of Energy-Efficient Non-Suicidal THi4s - a Next-Gen Synbio Strategy

**Conference:** International Conference on Plant Synthetic Biology and Bioengineering

**Year:** 2018

**Proceeding:** 2nd International Conference on Plant Synthetic Biology, Bioengineering, and Biotechnology

**Group:** General Submissions

**Session:** Transgenic Approaches for Improving Traits and Performance of Crops

**Time:** Saturday, December 1, 2018 - 12:00pm-12:25pm

## Authors:

Joshi, J., Western University
Hanson, A. D., University of Florida
Reisch, C. R., University of Florida

Around half the carbon captured by photosynthesis is lost by respiration and half of that respiratory carbon loss is driven by protein turnover and other maintenance processes. Scaling back maintenance costs by slowing the turnover rates of fast-turnover proteins could therefore save carbon and energy for allocation to biomass yield. Until now, SynBio approaches to increasing crop biomass yield have overlooked carbon losses and have instead focused on carbon gain (photosynthesis) because molecular targets can be defined; molecular targets to reduce carbon loss have been elusive. The situation has changed thanks to advances in proteomics that enable measurement of turnover rates of individual proteins. For instance, the thiamin synthesis enzyme THI4 is now known to turn over extremely fast, and bioenergetic calculations indicate that this turnover could account for up to 5% of maintenance costs, or 1% of carbon fixed by photosynthesis. An efficient THI4 substitute is therefore a Next-Gen SynBio target for crop improvement. Canonical plant THI4s are suicide enzymes that use an active-site cysteine residue as sulfur donor and so are irreversibly inactivated after a single reaction cycle. Certain anaerobic prokaryotes have THI4s that use sulfide as sulfur donor and so are energy-efficient catalysts; these enzymes have histidine or another residue in place of the active-site cysteine. Screening such THI4s by functional complementation of an *Escherichia coli* thiamin auxotroph identified some that can function in aerobic conditions similar to those in plant cells. However, the low prevailing sulfide levels *in planta* may limit the activity of such sulfide-utilizing THI4s. Applying directed evolution, e.g., with the EvolvR system, can potentially modify candidate THI4s such as that of *Thermovibrio ammonificans* to function efficiently in aerobic, low-sulfide conditions. Combining parts-prospecting for energy-efficient SynBio target THI4s with a directed evolution approach is thus an innovative path to boost crop productivity.

# Synthetic Design of New Plant Oil Traits for Oilseed Crops As Sustainable and High-Quality Feedstock

**Conference:** International Conference on Plant Synthetic Biology and Bioengineering

**Year:** 2018

**Proceeding:** 2nd International Conference on Plant Synthetic Biology, Bioengineering, and Biotechnology

**Group:** General Submissions

**Session:** Transgenic Approaches for Improving Traits and Performance of Crops

**Time:** Saturday, December 1, 2018 - 12:25pm-12:50pm

## Authors:

Kim, H., University of Nebraska-Lincoln
Cahoon, E., University of Nebraska
Clemente, T., University of Nebraska

Using soy products as fish food ingredients is sustainable and viable for fish aquaculture. However, the current soy-based feed lacks EPA/DHA omega-3 fatty acids and oil-based feed components, like astaxanthin, needed to obtain the pink-color of some fish meat. Because of these deficiencies, soy-based aquaculture feed currently requires supplementation with fish oil and high-priced astaxanthin, particularly for farm-raised fishes like salmon and trout. In addition, oils with high very-long-chain polyunsaturated fatty acids content are easily oxidized, which limits the shelf life of fish due to the development of off-flavors and odors. For that reason, a high vitamin E antioxidant trait is needed to provide oxidative stability. Conventional plant breeding and biotechnology approaches typically target only one or a small number of traits at a time. In contrast, the emerging discipline of synthetic biology offers tools for making step changes in crop improvement by enabling integration of many trait genes into the crop genome in a single genetic transformation. The goal of our research is the stacking of multiple oil traits: EPA/DHA, astaxanthin, and vitamin E, to obtain a single soybean line with optimized aquaculture feed value. To produce EPA, five transgenes including desaturase and elongase from diverse organisms were synthesized with barley *HGGT*, which is involved in the production of vitamin E. For astaxanthin production, *Î*$^{2}$-ring oxygenase and astaxanthin synthase from *Adonis aestivalis* were used. An 8-transgene expression vector with seed-specific promoters was constructed and transformed into soybean. T3 seeds successfully produce EPA, vitamin E, and astaxanthin. To evaluate the upper limit of transgene numbers that can be introduced into plant from one construct, we also constructed ten modules with multi trait genes to generate improved sustainability and biofuel and industrial quality traits. These trait targets include genes for heat tolerance, seed size, oil content, oil quality, and protein quality.

# The GRAIN Platform: Identification of Gene Targets for Improving Crop Yield

**Conference:** International Conference on Plant Synthetic Biology and Bioengineering

**Year:** 2018

**Proceeding:** 2nd International Conference on Plant Synthetic Biology, Bioengineering, and Biotechnology

**Group:** General Submissions

**Session:** Transgenic Approaches for Improving Traits and Performance of Crops

**Time:** Saturday, December 1, 2018 - 12:50pm-1:15pm

## Authors:

Skraly, F. A., Yield10 Bioscience
Snell, K. D., Yield10 Bioscience
Ambavaram, M. M. R., Yield10 Bioscience
Malik, M. R., Metabolix Oilseeds
Peoples, O. P., Yield10 Bioscience

Achieving step changes in the seed yield of major crops is not likely to come about with single genetic manipulations, but rather with combinations of changes that are conceived through modeling, experimental iteration, and logical reasoning. Drawing upon its many years of metabolic engineering and modeling expertise, Yield10 is developing the GRAIN (Gene Ranking Artificial Intelligence Network) platform, which encompasses metabolic and regulatory modeling, along with a robust transformation and field trial system. Combinations of genetic changes for increased yield may encompass both traditional plant breeding and molecular breeding tools such as editing, transgenics, and rearrangements of native genetic elements. The GRAIN platform seeks to optimize combinations of these changes by utilizing stoichiometric, kinetic, thermodynamic, and co-expression analyses to narrow the list of actionable targets from many thousands to relatively few, providing plant scientists with manageably small libraries of manipulations to test. The iterative nature of GRAIN is enabled by using camelina as a model crop, because of its ease of transformation and regeneration, its rapid and compact growth, and the relatively few regulatory hurdles associated with field trials. Using this approach, Yield10 is identifying and de-risking yield traits for major agricultural crops.

# Transcriptional Rewiring: Editing Plant Stess Response Networks

**Conference:** International Conference on Plant Synthetic Biology and Bioengineering

**Year:** 2018

**Proceeding:** 2nd International Conference on Plant Synthetic Biology, Bioengineering, and Biotechnology

**Group:** General Submissions

**Session:** Transgenic Approaches for Improving Traits and Performance of Crops

**Time:** Saturday, December 1, 2018 - 1:15pm-1:40pm

## Authors:

Windram, O., Imperial College London

Plant defence responses are modulated by substantial transcriptional reprogramming, up to 40% of the genome can be differentially expressed following pathogen challenge. This group of differentially expressed genes includes hundreds of transcription factors suggesting a complex transcriptional network regulates plant stress response. We have previously shown that transcriptional rewiring can be used to generate novel phenotypic diversity that can help adapt yeast to stresses associated with heterologous expression constructs. Transcriptional rewiring involves the fusion of transcription factor coding sequences with non-natural promoter sequences which in turn can have pleotropic cascade effects on downstream target cell pathways. Here I will show how we are using a combination of Synthetic and Systems Biology, inspired by our studies in yeast, to select plant regulators for targeted rewiring approaches that seek to enhance plant tolerance to pathogens.

# Genetic Engineering for Enhancing Abiotic Stress Tolerance in Canola (Brassica napus)

**Conference:** International Conference on Plant Synthetic Biology and Bioengineering

**Year:** 2018

**Proceeding:** 2nd International Conference on Plant Synthetic Biology, Bioengineering, and Biotechnology

**Group:** General Submissions

**Session:** Transgenic Approaches for Improving Traits and Performance of Crops

**Time:** Saturday, December 1, 2018 - 1:40pm-2:05pm

## Authors:

Bhalla, P., The University of Melbourne
Tabasum, A., The University of Melbourne
Singh, M., The University of Melbourne

Canola is the second largest oilseed crop worldwide used for human and animal feed. Canola cultivation has grown rapidly over the past five years owing to a high demand for canola oil and meal and is now Australiaâ€™s third largest broad-acre crop. Australia is the world's second largest exporter of canola seed. With rising global demand for canola for food and non-food applications, its production is expected to increase by 40% by the year 2025. However, impending global climatic changes are predicted to hamper crop productivity. Salinity, drought and high temperatures are significant environmental factors that limit agricultural yields. In this context, genetic improvement of crops for abiotic stress tolerance is vital to maintaining our food supply.

Helicases, an important class of DEAD-box protein family, are primarily known to unwind duplex nucleic acids to perform many housekeeping activities. These highly conserved enzymes play an essential role in several cellular processes including RNA metabolism and regulation of gene expressions. The development of abiotic stress tolerant canola lines by heterologous overexpression of a DNA/RNA helicase gene will be presented.

# Cotton Fiber Quality Improvement By Introducing Sucrose Synthase (SuS) Gene into Gossypium hirsutum L

**Conference:** International Conference on Plant Synthetic Biology and Bioengineering

**Year:** 2018

**Proceeding:** 2nd International Conference on Plant Synthetic Biology, Bioengineering, and Biotechnology

**Group:** General Submissions

**Session:** Transgenic Approaches for Improving Traits and Performance of Crops

**Time:** Saturday, December 1, 2018 - 2:05pm-2:30pm

## Authors:

Shahid, A. A., Centre of Excellence in Molecular Biology, University of the Punjab
Ahmed, M., Centre of Excellence in Molecular Biology, University of the Punjab

The demand for long staple fiber having better strength and length is increasing with the introduction of modern spinning and weaving industry in Pakistan. Research on exploring the molecular events in developing fibers reported several genes that play decisive role in different stages of fiber development. Sucrose synthase (SuS) helps in metabolism of sucrose in a plant cell, in cotton fiber it controls a bidirectional reaction but favorably transforms sucrose and UDP into fructose and UDP-glucose. UDP-glucose (UDPG) is a nucleotide sugar act as a donor for glucose residue in many glycosylation reactions and is essential for the cytosolic formation of sucrose and involved in the synthesis of cell wall cellulose. The study was focused on successful*Agrobacterium* mediated stable transformation of SuS gene in pCAMBIA 1301 into cotton under a CaMV35S promoter. Integration and expression of the gene were confirmed by PCR, assay, and real time PCR. Young leaves of SuS overexpressing lines showed increased total soluble sugars and plant biomass as compared to non-transgenic control plants. Cellulose contents from fiber were significantly increased. SEM analysis revealed that fibers from transgenic cotton were highly spiral and fiber twist number increased per unit length when compared with control. Morphological data from field plants showed that transgenic plants performed better in field conditions. Incorporation of genes related to cotton fiber length and quality can provide new avenues for fiber improvement. The utilization of this technology would provide an efficient import substitution and sustained production of long staple fiber in Pakistan to fulfill the industrial requirements.